T0378286

SPOTLIGHT ON OUR FUTURE

ENDANGERED SPECIES AND OUR FUTURE

SABRINA ADAMS

NEW YORK

Published in 2022 by The Rosen Publishing Group, Inc.
29 East 21st Street, New York, NY 10010

First Edition

Editor: Theresa Emminizer
Book Design: Michael Flynn

Photo Credits: Cover Richard Whitcombe/Shutterstock.com; (series background) jessicahyde/Shutterstock.com; p. 4 WishnclickS/Shutterstock.com; p. 5 Sushaaa/Shutterstock.com; p. 6 Nenov Brothers Images/Shutterstock.com; p. 7 Shannon Dunaway/Shutterstock.com; p. 9 (cicada) Toru Kimura/Shutterstock.com; p. 9 (deer) marcin jucha/Shutterstock.com; p. 9 (trout) Harel Bartik/Shutterstock.com; p. 10 Ikordela/Shutterstock.com; p. 11 Fiona Ayerst/Shutterstock.com; p. 12 Lukasz Pawel/Shutterstock.com; p. 13 Popperfoto/Getty Images; p. 15 Herschel Hoffmeyer/Shutterstock.com; p. 16 Saul Loeb/AFP/Getty Images; p. 17 https://commons.wikimedia.org/wiki/Category:Phocoena_sinus#/media/File:Vaquita4_Olson_NOAA.jpg; p. 19 TigerStock's/Shutterstock.com; p. 21 (finches) UniversalImagesGroup/Getty Images; p. 21 (Darwin) GraphicaArtis/Archive Photos/Getty Images; p. 22 Rich Carey/Shutterstock.com; p. 23 John Carnemolla/Shutterstock.com; p. 25 Dmytro Zinkevych/Shutterstock.com; p. 26 Portland Press Herald/Getty Images; p. 27 RPBaiao/Shutterstock.com; p. 29 Bloomberg/Getty Images.

Cataloging-in-Publication Data

Names: Adams, Sabrina.
Title: Endangered species and our future / Sabrina Adams.
Description: New York : Rosen Publishing, 2022. | Series: Spotlight on our future | Includes glossary and index.
Identifiers: ISBN 9781725323896 (pbk.) | ISBN 9781725323926 (library bound) | ISBN 9781725323902 (6 pack)
Subjects: LCSH: Endangered species--Juvenile literature. | Nature--Effect of human beings on--Juvenile literature. | Wildlife conservation--Juvenile literature.
Classification: LCC QL83.A33 2022 | DDC 591.68--dc23

Manufactured in the United States of America

CPSIA Compliance Information: Batch #CSPK22. For further information contact Rosen Publishing, New York, New York at 1-800-237-9932.

CONTENTS

CHAPTER ONE

SMALL CREATURES, BIG PROBLEMS

Between 2006 and 2007, beekeepers around the world started to notice that bees were disappearing. Scientists called this problem "colony collapse disorder." Colony collapse disorder happens when the worker bee population of a bee colony, or group, disappears, while the queen and the young bees remain. A bee colony without worker bees will die.

Worker bees gather food, make honey, and take care of the young. Without them, the colony will die.

Bees **pollinate** plants. People and animals need these plants for food, homes, and more. Without bees, it may become harder for farmers to grow crops. This could lead to famine, or a shortage of food. No matter how small, all species, or kinds, of animals and plants on Earth are important. Many species depend on each other to survive. If one species dies out, it's likely that other species will follow.

CHAPTER TWO

INTERDEPENDENCE OF SPECIES

Interdependence means that species depend on one another to survive. Even humans need other species to survive. In early times, people gathered food from plants and trees. Later, people began growing these plants and trees on their own.

People also use plants for medicine. The aloe plant can help heal burns. Scientists are using plants to find cures for diseases, or illnesses, such as cancer. Some plants come from areas in danger of being destroyed, or ruined. If these plants go extinct, or die out completely, we'll never be able to find out what they can do.

ALOE PLANT

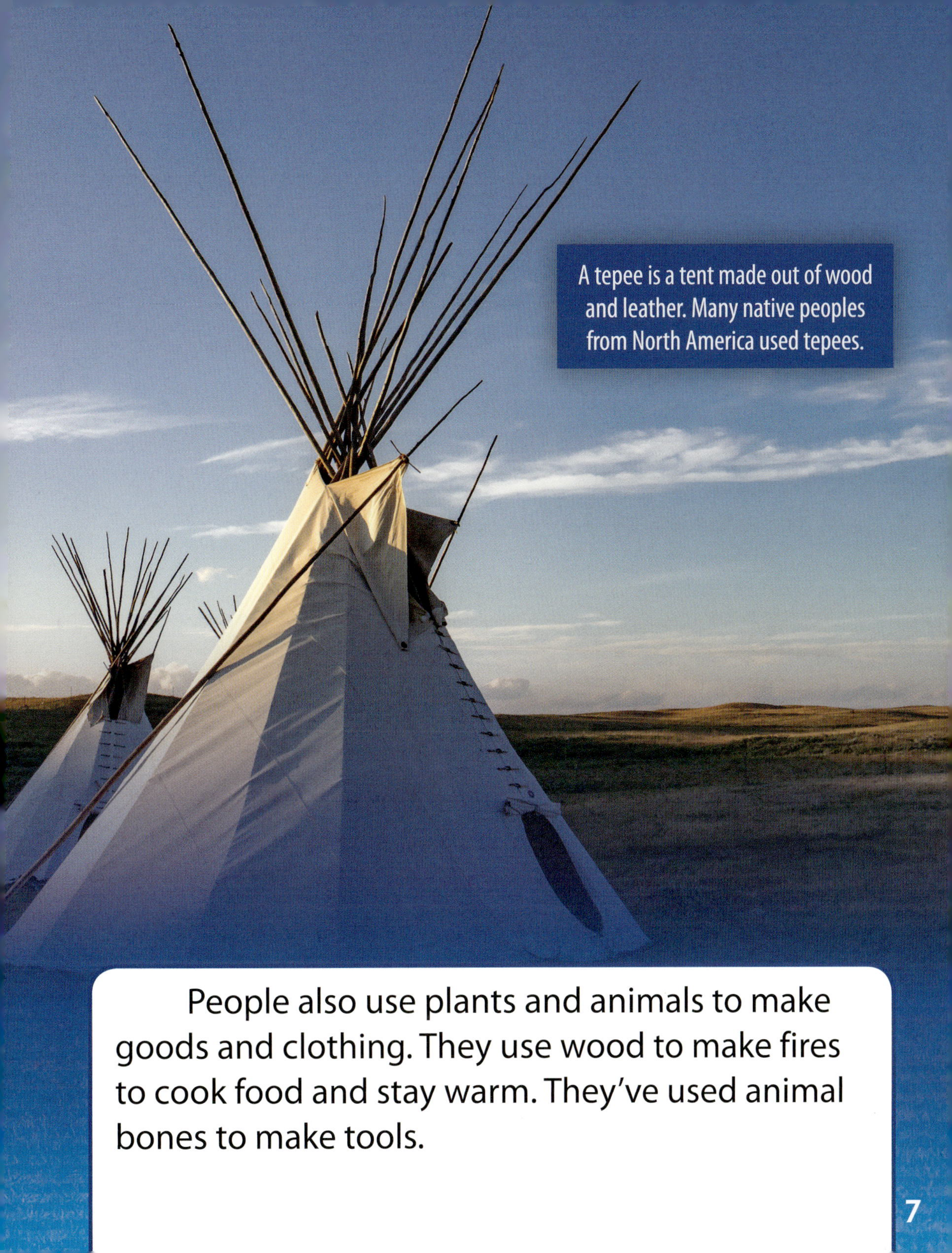

A tepee is a tent made out of wood and leather. Many native peoples from North America used tepees.

People also use plants and animals to make goods and clothing. They use wood to make fires to cook food and stay warm. They've used animal bones to make tools.

CHAPTER THREE

PART OF A COMMUNITY

A community is a group of people who share a way of life, values, or religious beliefs. Communities often share an area, such as a country, city, or neighborhood.

People in a school, workplace, or religious group can also be communities. For example, you and your neighbor might go to different schools. You both belong to one community where you live, but you also belong to different school communities as well.

An ecological community is a group of all the plant or animal species living in the same place. An ecological community is similar to a neighborhood community in this way. It depends on a balance of nature. This means that what affects one species will also affect many of the other species in the community. Interactions between the different species maintain the balance of a community.

An ecological community may also be called a biome. It may also be called an ecosystem.

CHAPTER FOUR

SPECIES INTERACTIONS

An ecological community has four main species interactions: mutualism, commensalism, competition, and predation. In mutualism, both species benefit from one another. One example is the relationship between bees and plants because both species benefit from pollination.

A remora fish fixes itself to larger fish and sea animals. Remoras feed on food left over after the host animal feeds. This is an example of commensalism.

In commensalism, one species benefits from another by gaining food, living space, or something else. The other species is neither helped nor harmed. One example is when tree frogs use plants to hide from predators.

Competition happens when two species compete for a limited resource such as food, water, or living space. With competition, both species can be negatively, or badly, affected. One example is when lions and cheetahs hunt the same prey.

In predation, one species benefits while the other is negatively affected. Predation occurs when one species uses another as a food source.

CHAPTER FIVE

WHEN SPECIES DISAPPEAR

When all the members of a species disappear, it's called extinction. Extinction isn't uncommon. It's a part of the natural processes of life on Earth.

When a species becomes extinct, it's gone for good. While this sounds terrible, extinction sometimes helps life on the planet.

Extinction can happen for different reasons. Sometimes a small change in the **environment** can cause extinction of one species. Sometimes species evolve, or change over time. They adapt, or change to live better in their environment.

An asteroid hitting Earth could lead to episodic extinction.

The thylacine was a large meat-eating animal from Australia, Tasmania, and New Guinea. The species went extinct when the last thylacine died in 1936.

When a slow, natural extinction happens, it's called background extinction. When extinction happens suddenly, it's called episodic extinction. This occurs when there are fast, major changes in the environment. An asteroid hitting our planet could result in a mass extinction of all or most species from an area.

CHAPTER SIX

MASS EXTINCTIONS IN HISTORY

Most animals that have ever existed on Earth have become extinct. **Catastrophic** environmental changes caused many of these extinctions. There have been five mass extinctions in Earth's past.

- The Ordovician-Silurian extinction—about 440 million years ago. Likely caused by continents shifting, which changed the climate.
- The Devonian mass extinction—about 375 million years ago. Likely caused by climate change and less oxygen in the oceans. **Meteor** strikes and **volcanic eruptions** may have contributed to this.
- The Permian mass extinction—about 250 million years ago. Likely caused by climate change, volcanic activity, and asteroid strikes.
- The Triassic-Jurassic mass extinction—about 200 million years ago. Likely caused by volcanic activity, climate change, and changing sea levels.
- The K-T mass extinction—about 65 million years ago. Likely caused by extreme meteor activity, which killed the dinosaurs.

When the environment changed 65 million years ago, dinosaurs were unable to adapt and went extinct.

CHAPTER SEVEN

MASS EXTINCTION TODAY

Some scientists believe that a sixth mass extinction is happening right now. Unlike past mass extinctions, humans might be causing this one. Usually about one to five species go extinct per year. However, human activity is causing species to go extinct faster. Scientists think that between 150 and 200 plant and animal species are going extinct every day.

The vaquita is the most endangered ocean **mammal** on Earth. Vaquitas often get caught in fishing nets and die. As of 2019, there were fewer than 20 vaquitas in the wild.

The extinction rate is now 1,000 times greater than that of a natural extinction. Human-caused pollution from **fossil fuel** use and the destruction many natural **habitats** are causing this.

There are ways to save species from extinction. Scientists now pay attention to what species are at greatest risk for extinction. There are about 1,800 species of plants and animals listed as endangered, or at serious risk of extinction, around the world.

CHAPTER EIGHT

HABITAT LOSS

There are two main ways a species can become endangered. One way is through the loss of its habitat. Habitats give species food, shelter, and space for reproduction. These habitats can change over time through events such as volcanic eruptions and climate change.

Sometimes, a species will adapt to habitat changes. Other times, habitat changes cause a species to go extinct. About 65 million years ago, Earth's climate cooled a great deal. The dinosaurs were unable to adapt and became extinct.

People also cause changes to habitats. Farming and logging have destroyed large parts of the Amazon rain forest. As human populations rise and people build more towns and cities, animal habitats are shrinking. Increased contact between people and wildlife can also lead to conflicts between species.

The California mountain lion has lost a lot of its habitat due to people. Destroying habitats puts species at risk of extinction.

CHAPTER NINE

GENETIC VARIATION

Loss of **genetic** variation, or difference, can also cause a species to become endangered. Examples of genetic variation are everywhere. Some people might have brown eyes while others have blue eyes. These differences happen because of the different ways that our genes combine.

Differences in a species allow it to adapt to changes. A large population of a species will have a large amount of genetic variation. Variations happen in one of three ways: mutations, or changes in the **DNA** of an organism; gene flow, or genes moving from one population of a species to another; or reproduction creating new combinations of genes in the species.

Human activities such as overhunting have caused some species populations to drop too low. That can lead to loss of genetic variation.

Charles Darwin, an English naturalist, found 15 species of finches on the Galápagos Islands. Each had a different beak depending on its diet. The finches evolved from a common species, in part through genetic variation.

CHARLES
DARWIN

CHAPTER TEN

HUMAN ACTIVITY

Humans affect Earth's species both in large and in small and hard-to-see ways. Wetlands are habitats that can include lakes, ponds, rivers, swamps, and marshes. Humans have drained many wetlands in Europe and North America so they can build on the land. Changing these habitats has negatively affected many species of water birds.

Logging leads to habitat destruction and puts species in danger.

The eastern barred bandicoot of Australia is endangered due to the destruction of its habitat. It also must compete with invasive species such as rabbits and deer.

Humans have also introduced invasive species to habitats around the world. An invasive species is a plant or animal that's not native to the habitat. Invasive species can take over a habitat, competing with other species for food, water, or space.

Pollution also impacts habitats in a negative way. Factory smoke can release poisonous chemicals into the air. This can also harm the water cycle, making polluted rain.

CHAPTER ELEVEN

MAKING A DIFFERENCE

How can people keep species from becoming endangered or extinct? Start by educating yourself! Talk to your family and friends about the endangered species in your area. Visit wildlife in local, state, or national parks. You could even **volunteer** at a nature center or talk to the people who work there.

Next, look around your home and see if it's wildlife friendly. Feed pets inside or in fenced-in areas. This will keep wild animals from eating food that may be harmful to them. Allow native plants to grow in your area. This provides food and shelter to native wildlife.

Make sure birdbaths are cleaned often to prevent the spread of disease. If possible, only fill bird feeders with food eaten by local bird species. The wrong kind of food may draw invasive bird species to your area.

Volunteering to help clean up parks can also help in your area.

Some young people are making big efforts to help endangered species. Hannah Testa of Georgia is a 17-year-old advocate for helping the planet. An advocate is someone who brings awareness to an important issue. Hannah was inspired to help when she learned about plastic pollution in the ocean. Now she spreads awareness about this problem and how it harms ocean wildlife.

Josiah Utsch, pictured here, was just 11 years old when he started Save the Nautilus. He works with his friend Ridgely Kelly and Dr. Peter Ward to teach people about at-risk species.

Olivia and Carter Ries of Georgia were just 7 and 8 years old when they created One More Generation. Their group educates people about endangered species.

Will and Matthew Gladstone are teenage brothers from Massachusetts. They created the Blue Feet Foundation. Their organization teaches people about a bird species called the blue-footed booby. The foundation raises money to protect wildlife by selling bright blue socks.

CHAPTER TWELVE

AN IMPORTANT LIST

In 1973, the United States Congress passed the Endangered Species Act (ESA). The act allows the federal government to protect endangered species, threatened species, and important habitats. The government has a list of protected plant and animal species in the United States and around the world. Species on the list are called listed species. The U.S. Fish & Wildlife Service and the National Oceanic and Atmospheric Administration enforce the act, or make sure it's followed.

The goal of the ESA is to help endangered and threatened species so that they can be removed from the list. It's illegal to harm, kill, or capture any species on the list. It's also illegal to interfere, or get in the way of, the other natural behaviors of a protected species.

In August 2019, the U.S. government under President Trump made changes to the Endangered Species Act. It weakened protections for endangered species and their habitats.

ENDANGERED SPECIES
Solar
Not Nuclear!
Ban Fracking Now
NO MORE
COAL OIL GAS FUELLING CLIMATE CHANGE

CHAPTER THIRTEEN

SUCCESS STORIES

The good news is that there are many endangered species success stories. Organizations such as the World Wildlife Federation and the National Audubon Society help save plants, animals, and habitats around the world.

Once, there were only 27 California condors left on the planet. The San Diego Wild Animal Park and the Los Angeles Zoo helped protect the bird and increase the species. Now there are hundreds of condors in the California skies.

The giant panda is another wildlife success story. Chinese and American scientists have worked together to bring pandas back from the edge of extinction.

Zoos and nature centers provide safe places for endangered and protected species. They can protect young animals until they are old enough to be returned to the wild. By working together, we can save endangered species!

GLOSSARY

catastrophic (kah-tuh-STRAH-fik) Related to or like a terrible disaster, or something that happens suddenly and causes much suffering and loss for many people.

DNA (DEE EN AY) A matter that carries genetic information in a plant or animal's cells.

environment (ihn-VY-ruhn-muht) The natural world around us.

fossil fuel (FAH-suhl FYOOL) A fuel—such as coal, oil, or natural gas—that is formed in the earth from dead plants or animals.

genetic (juh-NEH-tik) Referring to the parts of cells that control the appearance, growth, and other traits of a living thing.

habitat (HAA-buh-tat) The natural home for plants, animals, and other living things.

mammal (MAA-muhl) Any warm-blooded animal whose babies drink milk and whose body is covered with hair or fur.

meteor (MEE-tee-uhr) A piece of rock or metal that burns and glows as it falls from outer space to Earth.

pollinate (PAH-luh-nayt) To take pollen from one flower, plant, or tree to another.

volcanic eruption (vahl-KAA-nik ih-RUHP-shuhn) An event in which a volcano, or an opening in a planet's surface through which hot, liquid rock sometimes flows, sends out that rock and other substances in a sudden explosion.

volunteer (vah-luhn-TEER) To do something to help because you want to do it.

INDEX

PRIMARY SOURCE LIST

Page 13
Tazmanian "Zebra Wolf" Thylacinus in Washington, D.C. National Zoo. Photograph. Circa 1904.

Page 22
Darwin's finches or Galápagos finches. Illustration. Charles Darwin. 1845. *Journal of researches into the natural history and geology of the countries visited during the voyage of H.M.S. Beagle round the world, under the Command of Capt. Fitz Roy, R.N. 2d edition.*

Page 26
Josiah Utsch, 12 yrs. and Ridgely Kelly, 11 yrs. Photograph. Cape Elizabeth, Maine, December 20, 2012. Accessed through Getty Images.

WEBSITES

Due to the changing nature of Internet links, PowerKids Press has developed an online list of websites related to the subject of this book. This site is updated regularly. Please use this link to access the list: www.powerkidslinks.com/SOOF/endangeredspecies